Rock Crystal

Rock Crystal

The Magic Stone

compiled and edited by
Korra Deaver, Ph.D.

SAMUEL WEISER, INC.
York Beach, Maine

First published in 1985 by
Samuel Weiser, Inc.
Box 612
York Beach, Maine 03910

Revised edition 1986

ISBN 0-87728-577-2
Library of Congress Catalog Card Number: 85-50463

Cover art copyright © Jeanette Stobie, 1980
Reproduced by permission.

Typeset in 11 point Palatino by
Upright Type, Barrington, NH
Printed in the United States of America by
Interstate Book Manufacturers, Olathe, KS

Contents

*This book is dedicated
to independent thinkers everywhere
who dare to tread unconventional paths
and to explore the unknown, the unseen,
the unpopular and the forbidden.
May their kind increase!*

Preface

This book is a compendium of information about crystals that has been drawn from many sources. Where I knew it, I have tried to give credit to other authors, but through many years of study so much has been imprinted in my consciousness from the written word, lectures and workshops, that many of the original sources have been lost. Some common information has been found in more than one book or treatise, and has turned up in one or more seminar or workshop, and may be considered to belong to the realm of public consciousness. But to all researchers everywhere I extend a heartfelt thank-you for the information so freely and lovingly sent out so that all of humanity might benefit.

We have just begun to consider the many uses of the crystal. It will be quite some time before we completely understand its energy. Those who are involved in crystal research have experienced a great deal of personal growth and change on a personal level, for the search involves one's own mental, emotional, and spiritual energies.

We have made considerable advances on the road to spiritual attainment. Self-development and self-initiation are beginning to play a much more prominent part than has been true for several centuries. More and more people are no longer content to believe what they are told, but at last are desiring to know from their own experience. In this respect, the crystal can be used as a stepping-stone toward self-knowledge.

Crystal folk invite you to become involved with your own experiments, either in a group or alone. If you work with crystal exercises and meditations, you will probably have entirely different experiences than did the people who created them. Use caution and common sense in experimenting with anything new, and work with the sure knowledge that you are helping to take the crystal out of the realm of magic and folklore and are adding more to the common body of understanding, and that someday exploration with crystals may become a viable science in itself.

As to the future—who knows? Crystals have captured our fancy today just as they did when our ancestors used them. Many researchers are now working to recover the methods of the ancients and their apparent knowledge of crystal power generation and anti-gravity. It is reported that some are working to redevelop the Wilhelm Reich weather controller, which used crystals. New designs for space craft, space drives, intergalatic journeys and communications are being explored using crystal energies and magnetic grids. It boggles the mind to consider where this research may lead us.

Two future applications of crystals are described with some detail in *Exploring Atlantis, Vol. II,* by Rev. Dr. Frank Alper. Using the capacity of crystals to store energies, Dr. Alper postulates that crystals shall be used to store knowledge as in a computer. One can simply go into a library, move through the shelves of crystals until you find the area of knowledge you wish to absorb, and merely by holding the crystal, learn. This, says Dr. Alper, will become a reality in the future.[1]

---

[1]Reverend Dr. Frank Alper, *Exploring Atlantis,* Vol. II (Phoenix, AZ: Arizona Metaphysical Society, 1982), p. 67.

The other future application of crystal is in a health station. "Imagine," says Dr. Alper, "an individual enters a health station suffering from a chronic or terminal ailment. He is placed upon a machine. The machine analyzes and specifically pinpoints the quality and frequency of his vibrations. The computer analyzes them and prints out the exact degree and combination of color, sound and energy necessary to reharmonize the diseased organ in the body. Upon exposure to this triangulation of vibrations that is native to the perfection of this particular individual, with one treatment, all ailments are no longer present. The physical structure is returned to harmony and balance."[2]

Dr. Alper channels information from the past and the future as a conscious medium in tune with his own soul, "Adamis," and other energy complexes, some of which have had incarnations in ancient Atlantis.

People everywhere are experimenting with color, sound, magnets, orgone accumulators in a physical sense, and in combining mind energies with crystals for mental, emotional and physical healing. Some groups are using crystals to open and strengthen their telepathic connections with teachers, masters and healers from other dimensions by intuitive channeling. Others are working to open communication with plants and animals, and with the deva, angels of the mineral, vegetable and animal kingdoms.

It was only a few short years ago that the very ideas behind these experimentations would have been laughed to scorn. Today's workers couldn't be more serious. Their common intent is to create a new world of balance, harmony and love. This is a laudable goal from which no one is barred. Will you join us?

---

[2]Reverend Dr. Frank Alper, *Exploring Atlantis*, Vol. II (Phoenix, AZ: Arizona Metaphysical Society, 1982), p. 11. Reprinted by permission of the author.

Chapter One

Crystals in Science and Legend

Since the earliest times of recorded history, certain stones have been highly valued by humans as things of unparalled beauty—and of hidden power. Stones were considered lucky, unlucky, benevolent, malevolent, and some were believed to have the power to heal. The stories and legends of ancient peoples open up a fascinating world, revealing the extent of the influence that precious and semi-precious stones have had on the mystical imagination and of the power of enchantment they still may wield.

A great deal of interest is presently being generated about scientific and occult properties of the quartz crystals found in large quantities in Arkansas, California, and New York. Each area claims to have the most powerful and unique stones. Although, strictly speaking, the quartz family includes rose quartz, clear quartz, milky quartz, smoky quartz, amethyst, emerald, beryl, agate, carnelian, chalcedony, cat's eye (or Tiger Eye), onyx, sardonyx, and the various citrines, the stone capturing the popular attention at the moment is the clear rock crystal.

Rock crystal is transparent quartz, entirely devoid of the faintest trace of color, and, in its purest form is absolutely water-clear. Its hardness is 7 on the Mohs Scale of Hardness (topaz is 8; ruby is 9; and diamond—made of the hardest substance known—is 10). This scale was invented by a German minerologist, Friedrich Mohs, in 1922, and is used to determine the hardness of all gemstones. Hardness is determined by whether the particular mineral can scratch or be scratched by another mineral. The mineral which scratches another mineral is thereby determined to be harder than the mineral that is scratched.

Chemically, rock quartz is composed of silica (SiO_2). These stones crystallize into the rhombohedral system, and often form doubly terminated crystals of remarkable perfection and brilliancy. If the stone has one point only, the root half is often white and milky, rather than clear. Individual crystals can vary in size—some large stones weigh up to a ton, some are so tiny they can barely be discerned glinting in the sun.

According to John Rea: "Natural quartz crystals came out of the ground as you see them, without being faceted, broken off or altered by man, except inadvertently in mining. If the crystals are found in matrix they are kept that way. The breaks you can see were caused by earth changes ages ago, and most of them are at least partially healed by new crystal formation."[1] The unique molecular structure forms typical shapes of six-sided prisms acutely terminated by six planes: six-sided pyramids, not necessarily equal-sided. Doubly terminated stones are six-faced prisms with a six-faced pyramid at each end. Most areas have doubly terminated crystals which commonly have a cylinder of crystal between the two points which can range from a few centimeters to several inches in length. In New York State, unique stones called Herkimer crystals are found. They have two six-sided points end to end with no cylinder of crystal between, and they look rather like a double-ended toy top which could spin on either axis.

---

[1]From *The Quartz Crystal Story*, by John D. Rea, a pamphlet published by High Peak Crystal Co., Inc., 1272 Bear Mt. Ct., Boulder, CO 80303. Used by permission.

In mining operations quartz crystals are uncovered lining either or both sides of sandstone crevices from eight to one hundred feet deep in the ground. Some miners believe the crystals were formed by intense heat melting the silicon dioxide chemicals of which they are composed, forcing them into cracks and crevices where the chemicals reunited, condensed, solidified and became six-sided crystals. They will, when melted, resume a hexagonal shape as they cool. If cooled slowly, they form a single six-sided point. If cooled rapidly, they spangle out into a cluster of many points, all six-sided.

Crystal veterans can tell instantly the geographical origin of any particular crystal. They are mined from sandstone, pegmatite and clay-based soils into which weathered pegmatites have deposited their crystal. Sometimes you might find that sheets of crystal fall apart, leaving literally thousands of crystals up to a foot or more in length on a single sheet. According to John Rea, "Natural quartz crystals were formed in quartz veins and pockets that pushed into earlier sandstone deposits roughly 100 to 250 million years ago. The geometry of these formations is exquisite and intricate, and aligns with the magnetic fields of the earth. Crystals are different in each major deposit in the world, the main minable deposits being found in Arkansas, Brazil, and Madagascar.[2]

Kirlian photographs of various crystals show that the energy emanations are all different. Individual crystals produce identical patterns each time they are photographed, as long as the conditions remain the same, but each crystal has its own energy signature like fingerprints. That is why one should search for a crystal that has a special beneficent "feel" to it when choosing one's own personal crystal. The best results are obtained when your own energies and that of the crystal are harmonious.

As grown by nature, the crystal has piezoelectric properties, which means the crystal produces tiny amounts of electricity when squeezed and released, and it oscillates or vibrates with both a positive and a negative energy field.

---

[2]John D. Rea, *The Quartz Crystal Story*. Used by permission.

This precise frequency rate has made our modern world possible. Radio, satellite communications, telephone and television exist because of this fascinating energy activity. Recent solid state breakthroughs of quartz timepieces, radio, television, and other scientific advancements owe their existence directly to this crystal oscillation phenomenon, or knowledge derived from it.

Today rock crystals are used by science in computers, optical equipment, oscillators and sound resonators for radio transmitters, radar location equipment, and lenses for microscopes, cameras or any other instrument using a lens. It is used in any electronic device where incredibly high precision and sensitivity are required. Marcel Vogel, senior IBM research scientist in California, says in the future crystals will be used in healing and in thought photography, and in intergalactic and interdimensional communications. Crystal is so free from contamination that scientists use it to make their test tubes in which radioactive carbon age dating takes place.

The coming Age of Aquarius is prophesied to be a golden era of accomplishments unknown in recorded earth history. This may indeed be possible because of advanced oscillation communications allowing much higher levels of awareness and consciousness to be grasped by humankind. It is fitting that the mineral symbol for Aquarius be the rock crystal, for in its natural state, it is an excellent focus for expanded meditation, for polarity balancing and color therapy, to name a few. We will discuss some of these applications later.

Metaphysically, the rock crystal has been esteemed for many thousands of years as a heaven sent talisman. Since the early days of civilization, natural quartz crystals have carried the tradition of having mysterious, or even magical properties. One common ancient belief held that rock crystal was originally holy water that God poured out of Heaven. As the holy water drifted earthward, it became frozen into ice in outer space. This holy ice was then miraculously petrified by the guardian angels so that forevermore it would remain cold but would not melt and run away, thus preserving the holy water in solid form for the protection and blessing of mankind. Crystals are sometimes found containing small bubbles holding a few drops of water inside the stone, which moves about as the stone is turned. This could be part of the basis for believing the crystal is solidified ice. In luxurious ancient Rome, ladies carried crystal balls in their hands for cooling purposes during hot summer weather.

In Japan, the rock crystal is called the "perfect jewel." It is at once a symbol of purity and of the infinity of space, as well as symbolizing patience and perseverance. The Japanese believed that the smaller rock crystals were the congealed breath of the White Dragon, while the larger and more brilliant ones were said to be the saliva of the Violet Dragon. As the dragon was emblematic of the highest powers of creation, this indicates the esteem in which crystals were held. The name, Syusho, used both in China and Japan to designate rock crystal, also reflects the idea that rock crystal was ice which had been so long congealed that it could not be liquified.

The legend that rock crystal was a permanent form of ice might be interpreted as a memory of an aspect of sacred tradition. Occultly, rock crystal is assigned to Aquarius, the water-pourer, the giver of rain. According to Sir E. A. Wallace Budge, magicians of Australia and New Guinea use rock crystal for producing rain. Perhaps the original significance of the equivalence of rock crystal and ice was that they were both occult counterparts.

In his pamphlet entitled *The Quartz Crystal Story*, John Rea says that "the crystal was a sacred power object to the American Indians, Tibetan monks, Druid priests, and the priests and kings of Christian Europe as well. It was the stone of the White Light and of the First Ray, the Philosopher's Stone of the mineral kingdom, just as Saints and Holy Men were said to be the Philosopher's Stone of the human kingdom."[3]

Rock crystal is used in the mysticism of the Sufis and particularly in that of the Brahmins of India and the Lamas of Tibet. It is one of the seven precious substances of

---

[3]John D. Rea, *The Quartz Crystal Story*. Used by permission.

Buddhism, and in the Tibetan system the eastern region of heaven is said to be built of white crystal. One of the attributes of certain of the Hindu gods is the akshamala, or rosary of twenty-four crystals, representing the twenty-four Tattvas.

American Indians have long treasured crystal as a sacred stone. Medicine men used crystals as both a diagnostic aid in healing and as a means of communicating with the spirits. The presence of crystals with abraded edges in the mounds of Arkansas and elsewhere would lead to the inference that they were worn as charms and talismans, and having been used for such purpose, were probably buried with the dead as their property. Dr. Daniel G. Britton, in a paper on the folklore of Yucatan, says that the natives practiced witchcraft and sorcery, their wise men divining by means of a rock crystal, which was believed to exert great influence over the crops.

In the Scottish Highlands many crystal charm-stones were preserved for healing purposes. It was believed to cure hydrophobia, take away diseases of horned cattle, and to counteract the poison of infectious complaints. The usual practice was to apply to the patient—human or animal—water in which the stone had been dipped, along with certain magical incantations. A crystal set in silver and worn about the back was thought to be effective for diseases of the kidneys.

In Medieval Europe during the fifteenth and sixteenth centuries crystal was powdered for medicinal use, being given, mixed with wine, in cases of dysentery, diarrhea, colic, gout and to increase mother's milk. Pieces were held against the tongue for assuaging fever and to slake thirst. The Romans used quartz as a curative for glandular swelling, to reduce fevers and to relieve pain. In Japan rock crystal balls, mounted as charms, were worn to prevent dropsy and other wasting diseases.

As a magical talisman, crystal cannot be equalled. Ancient priests believed it to be a God-given force which defies all evil. It is believed to render negative energy impotent and harmless; it dissolves enchantments and spells, and destroys all black magic. White quartz will assist in elevating thoughts, while red quartz will aid positive action. Pink quartz is said to reduce wrinkles and to give the wearer a soft complexion. Yellow quartz or citrine will stimulate openness and will accelerate the awakening of the mind.

Crystal can be skillfully fashioned into powerful psychic tools such as meditation stones, crystal balls, aura feelers, tiny crystal pendulums, carved pendants, crosses and other talismanic pieces. Before the manufacture of glass, clear quartz was used for vases, cups and ornaments, with years being taken in the polishing and carving of especially clear specimens for the sculptor's art. Barring physical destruction, an object made of crystal could be said to endure forever as it neither ages, oxidizes or decays, and it is so hard that wear is seldom seen.

The crystal can be used by amateurs and professionals alike. It is not something that one can usually pick up and immediately use, and yet to each person there is a personalized sense of attunement with his own stone. Some crystals are warm to the touch, others cold. Some will absorb energy quickly, others more slowly, according to the ability of the user to harmonize his/her vibratory rate with that of the stone. It becomes different things for different persons for different needs. With this in mind, you can begin to experiment with your crystal in many, many ways. Find out which is your own particular gift and use it to your highest or ultimate goal. Know that it has an endless variety of uses and purposes, and it is up to you to define them. Experimentation has always been the basis for all scientific breakthroughs, and maybe you, my reader, will be the one who makes a new breakthrough in this field.

Even if the breakthrough is only in your own understanding of yourself-as-a-soul, as a Cosmic Being, your efforts will not have been in vain. According to John Rea, "It was once believed that quartz sand added to a garden brought the energy of the stars into the soil, and that something similar was true for man: quartz crystal brought the energy of the stars into the soul."[4] Each of us has the energies of the Universe within ourselves, as has the crystal, and it is well to learn to channel these to useful purpose, rather than to selfish or destructive ends. The crystal is of beneficial nature, not only to one's mind and body, but also to the spirit and soul. It symbolizes openness and clearness, and these are things one should strive for.

Each of the following chapters contain experiments you can try for yourself. Keep a journal in which to record the results of your own experimentations in this field. To get you started, we have created a mini-journal for you at the end of this book. From the keeping of careful records you may acquire enough information to write your own book some day, sharing what you have learned with others. Much is being learned about the technological uses of quartz, but not enough is known about the spiritual qualities that also exist within this magical stone.

---

[4]John D. Rea, *The Quartz Crystal Story*. Used by permission.

Chapter Two

Crystal Gazing

The art of scrying, or crystal gazing (in ancient terms called crystallomancy), has been practiced for ages by world leaders, seers, and conquerors. Not only does the crystal provide the necessary "clear deep," but its elemental essence possesses qualities which are distinctly stimulating to clairvoyance. It may even have a hypnotic effect.

Throughout the centuries the crystal has been the most popular of all oracles, with beryl, emerald, aquamarine, and clear rock crystal being the most favored stones for clairvoyance and seeing into the future. The colors of the beryl range from blue, through honey-yellow to actual transparency; the latter color resulting from the presence of peroxide of iron, while the green, and various shades of blue represent the effect of the protoxide of iron in varying quantities. The favorite shade of this crystal, as used by ancient seers, was that of the pale water-green beryl, or delicate aquamarine; water-green being astrologically considered as a color especially under the influence of the moon, which exerts a great magnetic energy.[1] All the various types of crystals contain oxide of iron, a substance with a strong affinity for magnetism. The ancient seers had strict injunctions to utilize the crystal only during the increase of the moon, which suggests that they believed there was a connecting link between the crystal and the unseen world of magnetism attracted to or accumulated in or around the crystal by the iron infused within it. The greater the increase of the moon, the greater the supply and accumulation of lunar magnetism in the crystal.

Two pieces of rock crystal clashed together in the dark show faint flashes of phosphorescent. Rub them together and they give a pale yellow light. The electrical poles of the crystal, first discovered by Baron Von Reichenbach, are at the ends of the three lateral axes. This force being most powerful at the opposite ends of a crystal, is therefore polar in nature.

Reichenbach published *Researches on Magnetism* in Germany in 1845, in which he maintained that every crystal exerts a specific action on nerves, which was especially remarkable in highly sensitive, or cataleptic persons. In certain diseases this force displays an adhesiveness towards the human hand, resembling that of iron to a magnet. Reichenbach claimed that a rock crystal when laid on the hand of a sensitive person instantly excited involuntary contraction, causing the hand to become clenched, and to grasp the crystal with a very strong spasm. For this new and specific force he suggested the name odylle, od, or odic force. Sensitive persons placed near crystals diffusing odylle feel comfortable only when the opposite poles of the same bodies are brought near. If the like poles are brought together, then very unpleasant sensations, sufficient to cause illness if left too long, were the result.

---

[1]William Thomas Fernie, M.D., *The Occult and Curative Powers of Precious Stones,* p. 204. Copyright © 1973 by Rudolf Steiner Publications. Source material used by permission of Harper & Row Publishers, Inc.

Reichenbach postulated "that there streams from the human eyes an efflux of magnetism as projected from its reservoir in the lesser brain (cerebellum), when the gaze is fixed intently upon a given point." This concentration is the method by which the gazer becomes enrapport with matters beyond his ordinary knowledge and in tune with the minds of others nearby, through the medium of the crystal.

The ancients also held it essential that there should be a concentration of unalloyed magnetism occupying the body of the operator by reason of purity of the sexual functions, believing (in common with Yogi masters) that sexual activity discharges energy from the body and diminishes the quantity and quality of natural magnetism. The search into the crystal being by the help of the blessed spirits, and open only to the pure from sin and to men of piety, humility, and charity, the occultists of old employed young boys before puberty and chaste innocent young virgins for crystal-gazing and clairvoyance, believing they made the quickest and sharpest seers; purity giving power in all magnetic and occult experiments. It was believed that men could not so readily be developed into seership as the female; but that they became superlatively powerful and correct, when so developed. Among women, virgins saw best; and next to them in order, widows.[2]

They also believed that the condition of the blood at the time of experimenting with the crystal was of great importance. There are approximately 38 grams of iron in the blood of the average person, while about one grain per day is taken into the body with food. This iron forms the coloring matter of the red blood corpuscles. The white (or colorless) blood corpuscles, which are much fewer in number than the red in a healthy body, are diminished by

---

[2]William Thomas Fernie, M.D., *The Occult and Curative Powers of Precious Stones,* pp. 205-208. Copyright © 1973 by Rudolf Steiner Publications. Source material used by permission of Harper & Row Publishers, Inc.

fasting, and increased by eating, which was of serious interest in connection with any prolonged abstinence from food prior to magnetic experiments with the crystal conducted by seers of the past.[3]

It was also believed that persons of dark, or very dark eyes, or very dark hair, eyes and skin were the most magnetic, which usually represented a dominantly bilious tendency or temperament with a larger amount of iron in the bile, being present as phosphate of iron. This pointed to the conclusion by the ancient seers that a certain chemical balance between the ferric and the oxygenic (i.e., magnetic) conditions of the blood and the bile was necessary towards developing the most perfect powers of concentration for crystal gazing and clairvoyance. To enhance this desired condition John Melville in his *Manual of Crystal-Gazing*, first published in London in 1897, suggested that the familiar English herb "Mugwort" (Artemisia vulgaris), or of the herb Succory (Cichorium intybus), would, if taken occasionally during the moon's increase, constitute for the crystal-gazer an aid toward the attainment of the best physical condition. It is an interesting fact that both of these plants are especially responsive to magnetic influence; their leaves, like the needle of a compass, invariably turn themselves toward the north. Furthermore, it was believed that these herbs act beneficially upon the generative system, thus influencing the function most closely allied to magnetic force.

To "charge" the stones with purity, magnetism, and spiritual qualities, set forms and rituals were employed. The ancient books were very strict in the types of formulas and rituals to be used, the times of day in which the stones could be used for scrying, even the kinds of persons who were allowed to use the crystal. Injunctions were given that

---

[3]William Thomas Fernie, M.D., *The Occult and Curative Powers of Precious Stones*, p. 206. Copyright © 1973 by Rudolf Steiner Publications. Source material used by permission of Harper & Row Publishers, Inc.

the moon must be in her increase; the crystal should be enclosed in a frame of ivory or ebony and stood upon the table; or, if simply held in the hand, it should lean away from the gazer, and should be held so that no reflections, or shadows appeared therein. Persons of a magnetic temperament, of the brunette type, dark eyed, brown skinned, and having dark hair, could charge the crystal more quickly, (but not more effectually) than those of the electric temperament, such as the blonde.

A full and complete ritual may be found in *Crystal Vision through Crystal Gazing*.[4] In this ritual the crystal is placed in a double circle of protection containing a triangle, a hexagram, a pentagram, and a Maltese cross, along with the name "Tetragrammaton," meaning Holy One, the Father, and the names of four principal angels who rule over the Sun, Moon, Venus and Mercury: Michael, Gabriel, Uriel, and Raphael. To each hour of the day and night were assigned the names of the angels and planets which rule that hour for each day of the week. The seer must first check the hour of the day and the week to determine which angel was available at that time, or wait until the right hour, if he would call upon a particular angel, before proceeding.

Several prayers and invocations were then intoned, certain promises made not to use the information for evil, and after certain rituals with rings, books, pentacles and perfumes, the angel thus invoked would appear within the crystal ball. The angel was then interrogated to determine his true name and office and true sign or character, in order to insure that a lying spirit had not usurped the angel's place within the sphere.[5]

At other times rituals were performed as an offering to St. Helen, "whose name was to be written on the crystal

---

[4]Frater Achad, *Crystal Vision Through Crystal Gazing* (Jacksonville, FL: Yoga Publication Society, 1923).

[5]From *Crystal Vision Through Crystal Gazing*, by Frater Achad, p. 75. Used by permission of Yoga Publication Society.

with olive oil, beneath a cross, likewise designed, while the operator turned himself eastward. A child, born in wedlock, and perfectly innocent was then to take the crystal in his hands, and the operator, kneeling behind him, was to repeat a prayer to St. Helen, that whatsoever he wished might become evident in the stone. Finally, the Saint herself would appear in the crystal, in an angelic form, and answer any question put to her." All this was to be done just at sunrise in fine clear weather.[6]

Thus was the mystic art of crystal-gazing in England and Europe in the early seventeenth century. The art became so generally accepted that it was avowedly employed as a means for detecting minor crimes and offences; until King James I—called "the wisest fool in Christiandom" by his contemporaries—passed laws making crystal-gazing a serious and punishable offence. Whereupon it fell into disuse and became almost extinct throughout the next two centuries, coming again into the forefront as a fashionable fad in the late 1800's.[7]

Today's seers are more pragmatic in their approach to the crystal, believing that there should be a naturalness in any exchange between all elements of consciousness on the planet, and that the elemental essences of the mineral, plant and animal kingdoms interact according to the purposes of each. There is more emphasis on thought as the control factor of whatever energies flow to and from the clairvoyant and the crystal, and the belief that all persons have equal opportunity to develop the psychic talents within themselves. There is less desire to compel the presence of spiritual beings in the crystal, and more desire for self-growth and self-understanding. Science and mysticism are not yet "buddies," but more and more scientifically oriented minds are becoming aware that

[6]William Thomas Fernie, M.D., *The Occult and Curative Powers of Precious Stones*, pp. 215-216. Copyright © 1973 by Rudolf Steiner Publications. Source material used by permission of Harper & Row Publishers, Inc.

[7]Fernie, *The Occult and Curative Powers of Precious Stones*, pp. 215-216.

something more than measurable energies are at work in the consciousness of all forms.

There are two types of crystal balls commonly in use. The favorite shape for divination is a sphere, ovoid or ellipsoid, set on a stand. There is also a much older psychic type of personal crystal ball which is not truly round. It is misshapen and irregular rather than perfect. It also is very smooth and highly polished, and it is often exactly the right size and weight to lie comfortably in the palm of the hand.

It is this personal type of crystal ball that many Bible scholars believe to have been the original "Urim and Thumin" of the old Testament. It has been suggested that the High Priest wore a thong suspended leather pouch under the twelve stone breastplate. In this leather pouch were two divining stones, a small polished piece of clear quartz crystal and a similar piece of smoky crystal. These two crystals represented day and night, guilty and innocent, left and right, yes and no, or any other opposites needed for the occasion.

Authentic crystal balls have always been exceedingly rare, so their use is necessarily restricted to the favored few. Most "crystal" balls on the market today are made of glass or plastic. Many are made of lead crystal, but this is still man-made glass, and neither substance has the oscillating positive and negative energy fields found in pure crystal. A genuine rock crystal ball, optically ground and polished, sells for many hundreds of dollars. At this writing I know of one seven inch sphere for sale at $30,000.[8]

One of the largest collections of crystal balls in the United States is at the Chicago Art Institute. It is recorded that a large pure quartz crystal ball over 18 inches in diameter was displayed at the 1904 World's Fair. Today it would be worth over $250,000.

Man-made acrylics, plexiglas and other types of plastics are frequently used to make "crystal balls" of

---

[8]Interested readers are welcome to write to me in care of the publisher about this.

surprising quality. They are very light in weight and it is easy to determine that they are not genuine crystal simply by picking them up and holding them in your hand. Lead crystal, on the other hand is more difficult to differentiate. The best man-made lead crystal balls are imported from Austria, and many 4-inch balls of good quality are available in occult shops for under $100. One can tell lead crystal from genuine rock crystal by holding it up to the light and looking through it for the flow lines that swirl around within the glass and were created when the ball was formed.

Rock crystal balls do not have flow lines within, but they may have tiny air bubbles, water bubbles, mineral stains, feathery cracks or other defects, called "veils." Of course, the more clear and perfect the stone, the higher the price it will command. The natural unpolished crystal point does have striation lines which go horizontally around the stone and can be seen when held up to the light.

Genuine rock crystal is usually heavier than lead crystal, but unless you have one of each to compare, testing by hand-held weight alone can be deceiving, for both kinds of crystal feel heavy when picked up. Lead crystal is standardized in weight according to size, but genuine crystal can vary in weight according to the amount of mineral deposits within the stone. Two rock crystals of equal size from different parts of the country can have a great variance in weight, one being substantially lighter than the other, when compared by feel.

Crystal gazing can be defined as the science of inhibiting normal outward consciousness by intense concentration on a polished sphere. When the five senses are thus drastically subdued, the psychic receptors can function without interference. Exactly why this alteration of consciousness works so well with crystal is not precisely understood, but many researchers believe there is an energy interchange between certain portions of the brain and the stone. Thought waves are energy very similar to radio waves. These waves of brain energy trigger the crystal into activity which in turn stimulates the dormant psychic centers to awaken and function. In a sense the crystal acts as a filtering antenna and amplifying reflector to the psychic centers.

The energy of the crystal varies according to the size and shape of the crystal being used. It can be changed and altered and this in itself can be quite an art. The crystal is able to tap the energies of the universe, with or without a particular person working with it, and yet it can be modified by human aura or touch.

The crystal will respond to the pressure of hand or fingers. It will actually generate a vitalizing current that the body's own electric field will be able to pick up. Focused on this energy in meditation, it can cause one to reach a point of infinity in which one is able to see all things clearly, for it is one's own mind projecting into the ball. It is not the crystal itself, but rather it is the fact that one can reach into this and become "at one" with all things through the third eye in the etheric body.

The quartz crystal can relate well to the heart center, also, but it responds best when using it with mental work, or the third eye center. Concentration on the stone seems to enable one to focus more clearly with the mind, rather than focusing upon any one particular form, thought, color or area. The mind can become a total blankness, a screen upon which images and thoughts are clearly reflected and magnified, or perhaps "accented" would be a better word. The blankness can be achieved, not because there is an absence, but because there is a "wholeness" about the stone.

Crystal can have a profound effect upon the crown chakra. Very potent in this area, it must be used with caution when experimenting with this center. The strong energy vibrations which emanate from it could be of a disruptive nature if one has not overcome the personal will, for when this chakra is opened, it must be made available to that which comes from the Highest Self, or the Soul, with the will quiescent and waiting to follow the Higher Guidance. If there is a sense of blockage when the High Center meets the resistance of the crystal, it can cause a painful experience, not only physically, but also create painful experiences in one's emotional relationships.

Learning How to Crystal Gaze

The following exercises in learning how to crystal gaze may be used equally well with any large reflective object, which can be a leaded glass or plexiglass ball, or clear colored gemstone, but since this book is about the properties of natural quartz crystal, we'll assume that you have a round or ovoid crystal sphere with a pure clear center not less than 2 inches in diameter. Of course, the larger the surface, the larger the area upon which a picture can be produced.

Easy naturalness is the key. One is merging one's consciousness with the elemental consciousness and purpose of the stone, which is a natural harmonizer and balancer. If one's intent is to bring about harmony and balance in one's life or that of another, then there is no conflict with the purpose of the stone, and seership comes easily and naturally with practice.

Among the many books and articles on crystal gazing the following thirteen steps seem to be universally accepted:

1. Practice is best carried out alone and in a plain room, as the degree of concentration is higher when one can sit quietly without distraction from mirrors, ornaments, pictures or glaring colors. Daylight, candlelight, moonlight, sunset, dawn, or drawn curtain shades seem to be the preference of the gazer rather than a necessity. Experiment to find the time which is best for you. The temperature should be comfortable and the light in the room should fall over your back, so that it is not reflected in the crystal.

2. You may prefer to hold the crystal in your hands, or you may wish to set it on a pedestal. In either case, a black or deep blue velvet or silk cloth should be placed between your hands and the sphere, (or the pedestal and the sphere) to prevent the distorted image of fingers or pedestal from showing through the clear stone. Adjust the cloth so that it cuts off any light or image reflection. The

cloth also helps to prevent a condensation of moisture, such as appears when hands are brought into contact with any cold surface. In the case of natural quartz crystal, however, the stone quickly warms and adapts to the energies of the gazer, and condensation disappears.

3. Wait at least an hour after eating a meal, as the energies used for digesting food may subtract from the energies used for mental development. Of course, avoid alcoholic beverages or drugs which dim or change the brain's clarity and control. Mental anxiety or ill health are not conducive to the desired end.

4. An inner peace is required. Do not try to visualize when you are upset or overly tired. Wait until you can begin with a calm, relaxed state of mind. Getting the brain into the alpha level is essential, as a deep meditative state of mind works best. If there is a royal road to crystal vision, its password is "Calmness, Patience and Perseverance."

5. Other aids to concentration which employ the five senses may be used. By gazing at the crystal you use your sense of sight, and there is less chance of being distracted by your surroundings. You may burn incense, employing your sense of smell, or you may hold the crystal which employs your sense of touch. Hearing should be muted, as it tends to bring in outside distractions. The solitude of the room itself is restful and aids in the mastery and control of your body and mind.

6. Begin your practice with an active visualization. Take a simple object, such as a silver spoon or a ballpoint pen. Hold it, feel it, look at it, concentrate on it, use all of your senses on it. Note such things as color, shape, size, texture, weight, etc. Close your eyes and visualize in your mind the object you have been studying. Then transfer the vision into the crystal. Try to see it in the crystal ball. Visions are normal to everyone. Think of something, anything, and it will appear in your mind's eye. Research shows there are two different types of brains—visualizers and non-

visualizers. More than 95% of the population belong to the visualizer-type, while only 5% make up the non-visualizers. If you among the 5%, then you will have to approach your crystal ball in a different way. It is *your* brain, and you alone can learn to use it most effectively.

It is well to note here that nothing is actually seen in the crystal. By concentrating on the crystal you create a meeting point for the imagination and the visualizing faculties, thus creating a vision which is actually in the brain itself. The transfer of the vision to the crystal works in the same way as the sense of sight. One actually sees with the back portion of the brain, but the faculty of sight seems to project the vision outward and we appear to see that which is external to ourselves. If you are a non-visualizer, become aware of how it is your brain does work, and project the outcome of the normal working of your brain into the crystal. You will achieve, for you, the same results as the visualizer.

If your visualizing faculty is weak due to lack of use, it can be enhanced by recalling past memories. Remember everything you did today, yesterday, last week. Recall the activity in full detail, until you can actually seem to see and feel yourself reliving the entire day or episode. Expand your memory back into the past to events of your childhood, to people you knew and loved. Recall vividly to memory features of a grandparent's face, for instance, or someone you knew and loved as a child. Project your memory pictures into the crystal.

Go from the known to the unknown. Try to imagine what something might look like that you've never seen before: Niagara Falls, the inside of a covered wagon, the Alaskan mountains. Ask a question. The mind will supply the answer.

7. Don't just look at the crystal—look into its depths—but don't stare. As you gaze into the crystal, avoid blinking your eyes more than necessary, yet don't strain to keep from blinking. The muscles of the eyelids will strengthen

with practice until an unwinking gaze can be easily maintained for twenty minutes to half an hour.

8. Concentrate on what you desire, with the expectation of seeing pictures. Don't let your eyes wander from the ball, nor let your attention relax from the subject in mind. Look intently into the depths of the crystal, concentrating on the subject that you wish to visualize. You will find that your eyes fall out of focus and you are then truly gazing into the heart of the crystal, and what is known as the "clear deep" will appear.

At first, spend no more than five to ten minutes in concentration upon the crystal. You can lengthen the time with later practice. For your first few tries you might have little or no success, but there are those who claim perfect visions during their first trial. Don't be discouraged, nor think your time wasted if you do not obtain visions during your first few sittings, or even if you never develop the ability. This is a method of mind development, and the concentration practice alone is worth the time and effort given to it. Concentration practice serves to strengthen the will and with that strengthening of will comes more power of choice and a richer life style. As with facility in any activity, sport or other skill, it usually takes time, patience and practice to develop the art of scrying, and to learn to use it effectively. The time spent will be well worth all your efforts, as you begin to achieve some guidance and dominion over your affairs.

You may be able to see things quite clearly one day and have no success another. This is quite normal for beginners, but practice will enable you to overcome this handicap, and some day you will find yourself in complete control of the mental faculties that function so effectively through the crystal ball.

Symptoms of Success: Just before a vision appears, the ball may cloud over as though a milkish gray mist were filling it, or a mist may appear as though it were interposed between your eyes and the crystal. This is called "clouding" and is the first sign of oncoming visions. Sometimes this clouding will appear to darken until it has formed a black, dark gray or deep indigo screen on which the visions will appear. Sometimes there may be small pinpoints of light, like tiny stars, glittering within, or the crystal alternately may appear and disappear as in a mist.

Sometimes the crystal seems to change color, starting with a dark red and proceeding through all the colors of the rainbow. The circles and rings seem to start at the center of the crystal and progress outward in concentric circles. The effect is similar to the disturbance seen in a pond when a pebble is thrown into it. The appearance of clouding (or the colors) is the first step to the actual vision, and may, or may not, continue for several sessions. A few more exercises are all that is needed at this point to strengthen your ability.

9. Breathing should be slow and deep while gazing, and it helps if you rest for five minutes with your eyes closed both before and afterwards. This will help avoid the fatigue that many clairvoyants experience when using the crystal. Remain passive, but alert, and do not try to see anything special at first. Then write on a slip of paper the question you want answered, or the future you wish to see; turn the paper over, and think no more about your question or problem, but continue passive and wait as before for what may chance to come.

10. In waiting for answers to questions or solutions to problems, it is important that you do not force yourself to see things in the crystal. The inner visions do not come by force, but by relaxation. Simply keep gazing into the center of the globe until reflections pass and melt away. Then comes the feeling of illimitable space around you, a sense of looking into a great void. With practice the pictures will become clearer and more brilliant. Sometimes as you gaze into the crystal no pictures will present themselves. Instead you will feel inside that you know the answers to your problem or question. This knowledge will be mental rather than visual. The subconscious can communicate with the conscious in many ways. It only requires faith and acceptance that the answers are already within you.

There are no set rules for explaining the meanings of symbols or interpreting the visions one sees. Quite often one finds he has complete understanding of the scene, though he may not be able to explain why. This, of course, is your own intuition at work, augmenting the psychic senses. Some people report seeing actual written messages in the crystal.

11. Try to visualize your dreams in the crystal. Visualization is a great assist to the memory, and if only for this reason crystal gazing is worth trying. Telepathy, or mind to mind communication, can be enhanced with the crystal. You can mentally request your spirit friends to try and make their presence known by vision in the crystal.

12. One word of caution: When you practice with your crystal ball in anticipating the future, don't expect problems, dangers, disaster. Expectation will create that which you expect. You are simply projecting ahead for the outcome of a current situation or present course of action. Your own emotions and expectations must be calm. Be willing to accept whatever is given to you from the subconscious storehouse of information, or the superconscious faculty of mystical knowledge, but don't project for any particular type of outcome.

Treat your mental excursions in the same way you would check the weather outside. If it is sunny, you will dress accordingly; if it is stormy, you wouldn't dress for warm weather. But in checking the weather, you neither expect sunshine or storm, you merely accept what is.

13. A reader must be responsible and take care how a precognition is worded to his client. A psychic, reading from the crystal ball for someone, may see a negative situation developing in the immediate future. He might say, "I see an accident about to happen," thereby making it seem destined and foreordained, and leaving the client apprehensive and worried. By adding his client's fears to his own vision, his very words may have solidified

substance so that the situation would, in fact, become unavoidable.

Instead, the reader might say something like, "On the job, take special care around machinery this week." The client, then alerted, will act in a positive way with sureness and deftness concerning his own safety, and by his very attitude expect to be in control of his environment, and quite probably avoid what might have been tragic consequences.

One of the best and most complete books on crystal gazing is *The Art of Crystal Gazing,* by Bevy Jaegers.[9] She provides a history of scrying as well as many complete exercises for developing crystal gazing. According to Jaegers, visions seen by scryers fall into six basic categories, as follows:

1. Imaginative visions, often called day-dreams, which come from the free, unrestrained thoughts of the person. For the constructive use of these, think only positive thoughts, if you want the crystal to amplify and enhance the positive, constructive energies within you.

2. Visions of forgotten events recalled from the memory. This establishes the link between the conscious and the subconscious mind. This type of visualization improves the memory and strengthens the mental faculties. One can use the ball to help develop one's memory by sitting quietly and looking into the depths of the sphere. Many details one cannot otherwise remember come clear in the crystal. Use the crystal to help remember where you mislaid something, or to help you remember facts or information that appear to be forgotten.

3. Visions of past occurrences, apparently unknown to the gazer. Usually these were incidents that actually did happen to the scryer, but they meant so little at the time they were ignored or unnoticed, but were recalled from the

---

[9]Bevy Jaegers, *The Art of Crystal Gazing* (Sappington, MO: Aries Productions, 1983).

subconscious storehouse of memory later, perhaps when the remembering made them more significant.

4. Visions of present events of which the gazer has knowledge. An important exercise in using the crystal is to have a friend visualize a familiar object such as a playing card or a tarot card in his or her crystal so you can try to pick up the same vision in your own crystal. This is more effective if emotion can be added to the vision being transferred.

5. Visions of present events of which the gazer has no knowledge. This is true mental telepathy, or thought transmission. Some believe these to be spirit messages, but they may also be a record of something happening to someone near or far away. The message received may have been intended, as in a previously set-up experiment, or may have been unconscious through an emotional bond between sender and receiver.

6. Visions of future events and predictions of the future. Often these visions are so filled with symbols, or seem so blurred in spots, that the event does not make any sense until it actually happens. Then the mind triggers back to the vision received in the crystal some weeks before. With constant practice the powers of the mind develop and soon one can accurately see the events of the future.

Taking time to classify the visions into these six categories, says Bevy, will serve as a guide, enabling you to watch your own development from fanciful day-dreams to accurate prediction of the future.[10]

Most people possess clairvoyant power in some degree. This power only needs concentration, practice, time and determination to blossom and become perfected. If you are already aware of your clairvoyance, it is possible to achieve wonderful results. Intelligent use of the crystal

---

[10]From *The Art of Crystal Gazing*, by Bevy Jaegers. Published by Aries Productions, Sappington, MO, 1983, pp. 4-6. Reprinted by permission of the author.

and its power can teach all of us its real value and influence.

The energies of the crystal are solidly founded in physical law. It is not necessary to believe in it for it to work. However, since it does function with the mental forces, it is possible for disbelief to have a lessening or blocking effect. Belief, or faith, can smooth the way for the crystal to work more effectively. Faith, passiveness, patience, continued study and practice should repay the crystal gazer in just proportion to the latent power within.

Always record what you see, as like a dream the crystal vision is easily forgotten. Writing down what you see not only impresses on the subconscious mind that you are serious in your desire to learn to scry, but also gives you something to return to, if the crystal vision does, indeed, turn out to be a prophecy of the future.

Record your own experimentation with the crystal in your journal as a record of your own spiritual journey into the deeper realms of mind and heart.

Chapter Three

Programming the Pendulum

The pendulum is one of the earliest methods of divination. It is a device for getting information from the inner self using some type of material suspended from a chain or a piece of string. Any material with some weight will do, but the most responsive material is rock crystal.

Choose a quartz crystal that seems to fit your own vibrations. It does not need to be perfectly clear. In fact, for pendulums, many people prefer a stone that is partially milky or with cracks or other refractions which give the stone "character." Its weight may be any size, but the smaller crystals are usually preferred, as they swing with more freedom than the larger stones.

When you've chosen your crystal, glue a bell cap on the end of it, opposite the point, and attach a chain or a thread or string. Or tie a loop around the end and add a length of string or chain so the crystal will dangle from the end. It is best if the crystal is balanced so the point hangs straight down. The pendulum swings by muscle movements governed by the parasympathetic nervous system of the subconscious mind. These are the same muscles that direct the involuntary actions of the body such as digestion, respiration and heart beat.

Some teachers train their students to command the pendulum to move as directed by the conscious mind: back and forth for "yes," a circle for "no," etc. In our researching, we've found that more people have success with the pendulum when cooperation by the subconscious is requested rather than ordered. The effect may be largely psychological, but who wants to argue with results? Occasionally a subconscious that has been dominated by a very strong will needs to be commanded to move in a prescribed direction, but for most people this is the procedure to program the pendulum to respond to the operator's direction:

1. Acting on the assumption that the subconscious area of the brain is a storehouse of memories of previous experiences, and that its behavior is learned behavior from those experiences, we first create the experience of a moving pendulum. Hold the chain of the pendulum between your thumb and forefinger at a comfortable height, probably three or four inches away from the stone, with your arm resting on the chair arm or table. Deliberately swing the pendulum first in a back and forth swing.

2. Stop the movement with your other hand, and then deliberately swing the pendulum in the opposite to and fro swing. Then move it in a clockwise circle, and then in a counter-clockwise circle, stopping the movement with your free hand between swings so that each action is separate and distinct. The order of the swings is not important, just that all four swings be deliberately introduced for recording in the subconscious mind.

3. Once the subconscious is alerted and aware of what you are talking about, you can then question it in this manner: "Which way will you swing to answer 'Yes' to my questions?" Wait and expect it to move in one of the four ways you have shown to it.

4. When the movement has occurred, ask each of these questions in turn: "Which way will you swing to answer

'No' to my questions?"; "Which way will you swing to answer 'I don't know' to my questions?"; and "Which way will you swing to answer 'I don't want to say'?". Always wait until a strong swing has been made. Then stop the movement from the previous question with either a command or with your free hand before going on to the next question.

In this way, the subconscious has an active part in seeking out the answers you request, rather than a submissive one as is the case when "ordered" to swing in a certain way to reply. The swing will always remain the same for each answer, and if you should forget which swing means what, simply ask the question again. The subconscious will gladly respond. You may question the pendulum aloud or mentally, it doesn't matter which.

The pendulum is an invaluable aid in searching the subconscious mind for hidden motivations, unknown conditioning or patterns of behavior, but that is an involved study which should be dealt with in a book by itself. The pendulum gives the usually unheard portions of the lower mind an opportunity to develop a communication with the external, or conscious aspect of mind.

Because the pendulum allows other portions of your mind to respond, the answers, "I don't know" and "I don't want to say" become extremely significant. The "I don't know" means that the answer to the question asked may lie in another region of the mind, or even in the Superconscious. In this case, you can ask the subconscious to go to the area of your mind or even to the area of God-Intelligence where the answer can be found, and to bring it back to you. Or you can raise your own level of aspiration upward and ask the same questions of the Superconsciousness aspect of yourself. Your desire is the motivation that brings the correct response from the proper area of consciousness.

The "I don't want to say" answer means you've touched upon some traumatic area deliberately or willfully hidden, perhaps some painful memory which you have literally told the subconscious mind that you don't want to think about anymore, a deeply repressed incident. Then psychological probing is needed to find and ask the proper questions that will elicit the information required.

A useful technique is to hold a crystal in the hand opposite the pendulum, and use the pendulum in the normal manner. Persons using this technique have reported as much as a tenfold increase in the power of the pendulum.

There are a number of ways in which the pendulum can be used to obtain information. The following list of exercises can be used in beginning to program the pendulum, to become familiar with how it will function for you. When you are really under pressure, looking for an answer, then your previous practice using these or other exercises will pay off.

Male and Female

1. The pendulum can provide very basic information for you such as determining the sex of an unborn child or unhatched chickens. Begin with holding the pendulum over the head or hand of a man or woman. It should swing in a circle over a woman, and in a straight line over a man, but let your pendulum choose its own swing. Occasionally we find a pendulum which wants to swing in the opposite way. Whichever it chooses, that is your answer always to male and female, wherever found.[1] Experimenting with the next two exercises will help to strengthen this aspect.

2. Place a number of objects on a table. Hold the pendulum over the objects, each one in turn. If it belongs

---

[1]There is an infrequent, but strange phenomenon that creates an apparent contradiction of the programmed male/female swings. You may hold the pendulum over a man's head or hand, for instance, and get a female swing, or over a woman and get a male swing. This may indicate there is an essentially female soul residing in the male body, or vice versa. If true, then further questions that can be answered "yes" or "no" might bring interesting new information. This is another area where vast research is needed.

to a woman, it will swing in a circle; if it belongs to a man, it will swing in a straight line (or the direction *your* pendulum has chosen for male and female). You can make this more difficult by sealing the articles in envelopes where they cannot be seen, so you cannot subconsciously influence the swing of the pendulum.

3. Use six photographs, three of women, three of men. Mix them up and place them face down on the table. Use the pendulum to discover which are the men and which are the women. Remix the photographs and try again.

Pendulum Detective Games

The following exercises will show how accurate the subconscious can be. Practicing first with using the powers of the subconscious will lead the way towards later use of the superconscious with equally sure results.

1. Choose five playing cards of the 8, 9, and 10 denominations. Four should be black and one red, or vice versa. Mix them up, or have someone do it for you, and place them face down on the table. Hold the pendulum over each card in turn, and ask it to find the lone card, red or black, by swinging in the "yes" motion.

Practice until you are always accurate in finding the single card. Then increase to seven cards with five black and two red, or five red and two black. When you have become expert in finding the two cards, increase the total number of cards until you are finally working with a whole deck at once, locating the red cards only. Then remix and use the same system to find the black ones. This demonstrates how important "intent" is when working with the energies of the crystal, since the pendulum will find black or red with equal ease, according to the intent and desire of the operator. Your will and your thought are the most important tools you have in influencing your own environment.

2. Secure three identical opaque containers. Fill one with water or any other substance, leaving the other two empty. Cap or cover them and have someone mix them up. Suspend the pendulum over each in turn, commanding it to swing in the "yes" motion over the one which contains the water.

3. Under one of three or four inverted teacups, place a metal object such as a ring, a watch or a coin. Use the pendulum to locate the object after the cups have been mixed up.

4. Write or print 35 to 50 common words on 3 x 5 cards, one word to a card. Use well known words with eight to ten letters each, and deliberately misspell about half of them. Shuffle the pack of cards and lay the deck face down on the far side of the table. Place one card at a time in front of you, face down, and ask the pendulum to answer "yes" or "no" to determine if the word on the other side of the card is misspelled.

Choosing Good and Bad

Vegetarians and people interested in health foods have used the pendulum to determine if certain foods contain pesticides or additives. Jewelers, art and coin collectors and dealers have used this system to detect phoney or counterfeit money, jewelry, coins or art. The following exercise will help you to program your own pendulum for this kind of information:

1. Allow the pendulum to choose whether it will circle for good and swing straight for bad, or if it wants to use its "yes" and "no" swings for good and bad (i.e., "yes" for good, and "no" for bad). Be sure your directives to the pendulum are clear, and its response will always be equally clear, once your communication system has been worked out.

2. Place three to five objects in a row, with one of the articles "bad." For example, several small batteries, one dead; some electric fuses, one burned out; a number of light bulbs, one non-working; a group of pocket mirrors face down, one broken; or salt one cup of water out of four or five in a row, and allow the pendulum to tell you which of the objects is "bad."

Finding Lost Objects or Persons

How often have you wished there was an easier way to find something that you have searched and searched for with no results? Here is one way you can immediately tap into the subconscious mind which never forgets where you have placed anything. Some psychic detectives also use the pendulum method for finding missing persons in conjunction with their local police department.

1. Suspend the pendulum over a map near the place in which you feel the missing person or object might be located. Ask the pendulum if you are correct or not. If the answer is "yes," move the pendulum to one of the compass points of the map. Ask it to swing towards the missing object or person. Draw a straight line exactly on the pendulum's swing.

Then move the pendulum to another compass point and repeat the request. When you have finished all four compass points, you should have four straight lines on the map, all converging at a certain point. Where these lines come together, you will find the missing target, whether it is a person or a lost object.

You may continue this exercise with more and more detailed maps of the pinpointed area, until you have precisely located the missing subject. This can be brought down to a certain house on a certain street.

2. If the pendulum indicates a lost object is still in your house, use the pendulum, asking "yes" and "no" questions,

to first locate the room in which the missing object lies. Then sketch the room, clearly indicating closets, shelves, drawers, etc. Then use the pendulum as in the previous exercise to locate the exact position of the missing object.

You may find the movements of the pendulum slow to get started, but you will become more and more accurate, the more often you do it. As with all other skills, practice is the key. Use the mini-journal to record your progress in using the pendulum. Keep track of both your hits and misses, and you'll soon find that the hits will begin to outnumber the misses. As your confidence mounts your number of hits will increase accordingly.

Chapter Four

Healing with Crystals

The crystal's piezoelectric properties can amplify, focus, store, transfer and transform energy. According to John Rea, "Natural quartz crystal was said to harmonize and align human energies (thoughts, emotions and consciousness) with the energies of the universe, and to make these greater energies accessible to man. In earlier times, when all things were thought to be conscious parts of a greater living consciousness, quartz crystal was thought to synchronize man's consciousness with that of the heavenly bodies and the hierarchies of angelic beings, visible and invisible."[1] Today's researchers who use crystals to amplify healing energies to alleviate human ills use different phraseology, but the inner concept is the same.

In this study, we do not recommend that you give up your own doctor to rely on crystal healing. Rather, working with a crystal should be an adjunct to the advice and treatment already being given. Someday a central control where all reported or documented information in this

---

[1]John D. Rea, *The Quartz Crystal Story,* a pamphlet published by High Peak Crystal Co., Inc., 1272 Bear Mt. Ct., Boulder, CO 80303. Used by permission.

field, including case histories, research and other newly discovered applications which might be tested by others, would help consolidate present theories into proven or disproven facts. At this time, many different types of experiments using meditation, massage, holistic health, polarity balancing, working with chakras, and other types of holistic therapy are being reported by people working with the crystal.

Choosing Your Crystal

The healing qualities of the crystal seem to be mainly an amplification of the energies of the one working with the stone. It amplifies the abilities of whomever is using it, changing to fit each particular person, and it can be used with most any illness. Of course, it is not a substitute for a doctor to set a broken bone, but it may be found that crystals can be helpful in relieving trauma, alleviating the pain of the injury, and in beginning immediate regeneration of affected tissues.

In most cases of healing diseased organs or tissues, the crystal is held in the area of disturbed energy on the body, usually represented by areas of pain or tensions, and as the energy field reharmonizes, so does the diseased tissue. There are no specific rules as to the exact distance the crystal is held from the body or the direction in which it is pointed. The patient himself must try various distances and orientations of the crystal until he feels an appropriate response, which may be heat, cold, a lessening of the pain or discomfort, a fluttering feeling around the distressed area, or simply a feeling of well-being. As always the best guidance is what "feels right."

Every healer will develop his/her own techniques, and there are an infinite number of combinations and variations. The results obtained from the different healing techniques are often subjective, and even if they work for some people, they may not work for another. Furthermore, the laws of karma must also be taken into consideration. No matter what the healing situation, there will be negative results if these laws have not been satisfied. Emotional need for the illness can also block healing. If being ill is the only way the patient can get attention, love or caring from his environment, then certainly he will continue to be ill, no matter how much healing is generated around him/her.

Using a crystal in each hand helps to amplify the healing forces that are flowing through the healer. Many people find that the most suitable shape for healing is the pyramid with six nearly equal sides, holding or strapping the pointed end against the palm with the largest or flat side facing outward or downward.

Pyramids formed with nearly equal sides have special properties. They focus and amplify through both the pyramid shape and the molecular structure of the crystal. The pyramid focuses energy through the pointed end. The spiral growth pattern of quartz causes energy to flow out of it in a spiral. When the two are combined, energy will flow from a pyramid point in a tight spiral beam which can be easily focused or directed.

Immediate results may or may not be experienced; remember, the stones symbolize patience and perseverance. You need to experiment with your own crystal to discover the best ways in which it will work for you.

Each crystal, according to size and shape, has a corresponding tone to which it can be aligned. Using tonal sounds along with the crystals during a healing exercise can set energy patterns to vibrating along specially directed lines. Practitioners in this art receive emotional benefits as a side bonus to the healing forces going to their clients.

There is a belief that only a gift crystal should be used for healing. Some people argue that since a gift of a crystal is a gift of love, the healings done through this special spiritual love vibration are more effective. Failing the necessary love donor, however, one can supply his own love vibration. The desire to be a healing channel in itself is a love vibration. If you are ready and want to use a crystal, by all means buy one. Take your time and select it with care. Touch and feel many crystals until you've found the one that seems to respond especially to you—or one to which you seem to especially respond.

It's lots of fun to go out and find your own crystal, but unfortunately, this is difficult for most people. Although it would be hard to find an area on the earth in which some variety of crystal is not found, good collecting areas are scarcer. Still, it might be surprising what is available in your neighborhood. You can find out by making inquiries at your local museum, geological society, or the geology or earth science department of a nearby university. Check the Yellow Pages for rock hound associations, mineral dealers and rock shops, who will have knowledge of, or access to, many good collecting locations. At the library or local bookstore you can also find guidebooks written for the collector which describe various collecting localities. If you live near a mining operation or quarry, there are excellent possibilities for finding crystals in the waste rock or tailings.

It may be tempting to have someone who supposedly "knows" more about crystals than you to choose a crystal for you, but to do so takes away from the opportunity to expand your own awareness and intuition.

The size and shape of the crystal you choose for your personal crystal should depend upon the "feel" you get from it when you hold it in your hand. It need not be a single point, but could be a doubly terminated crystal, or even an interesting piggy-back twin. One such type of twin looks like two crystals growing in the same space with double points of different sizes. The tetrahedral-shaped quartz molecules have a tendency to arrange themselves in spiral patterns. In the left-hand crystal the spirals turn to the left, but in the right-hand crystal, the spirals turn to the right. The right-hand crystal is identified by the presence of lozenge-shaped face(s) present to the right of the largest pyramid face of the crystal. If the face appears to the left of the pyramid face, it is a left-hand crystal. Most quartz crystals are a mixture of both. Researchers claim that different type energies are found according to the direction of the spiral.

Crystals amplify the entire energy field of the body, and if you wear a crystal over a particular area, it will focus more strongly in that area. If worn close to the neck, it will stimulate the thyroid and parathyroid glands, being especially helpful for respiratory problems such as congestion and sore throats.

Worn over the heart, crystal stimulates the thymus glands and increases the efficiency of the immune system for defense against disease. Wearing one over the solar plexus will cause a stimulation in total body energies, but also can increase your emotional field, which may not be so desirable.[2]

Experimentation with Kirlian cameras by Marcel Vogel, Dale Walker and others show that the direction in which the point is worn makes very little difference. There is an overall increase in the energy field, no matter which way it points, up, down, or sideways. None of the directions caused a decrease in the energy field. Pointing the crystal up channeled some of the energy into the upper chakras, causing some people to be so stimulated they felt they were leaving their bodies. When pointed horizontally, a small increase was noted in the energy in front of the body, with a large increase on the sides. When the point was down, there was a slight grounding effect as energy was directed toward the lower body, and tended to bring people back into their bodies and keep them better grounded and more in tune with the world. The researchers recommended that the point be worn up when meditating, praying, studying or taking tests, but worn downward the rest of the time.

---

[2]DaEl, *The Crystal Book* (Sunol, CA: The Crystal Company, 1983), p. 75. Copyright © by Dale Walker, 1983. Reprinted by permission of the author.

The physical body is only one of several bodies which make up the total being. These other bodies (esoteric philosophy says they are seven in number) are principally composed of various levels and states of energy interconnecting the mind, body and emotions. Crystals are responsive to these subtle energies, and healing begins within these etheric fields. Kirlian photography shows that if you hold a crystal in your left hand you will at least double the body's etheric fields. Since available energy can then be directed by thought, the possibilities are limitless in how that energy can be used to transform human lives, or even uplift the general atmospheric environment of the planet.

Crystals act as transformers and harmonizers of energy. Illness in the physical body is a reflection of disruption or disharmony of energies in the etheric bodies, and healing takes place when harmony is restored to the subtler bodies. The crystal acts as a focus of healing energy and healing intent, and thereby produces the appropriate energy.

The etheric body that is the most apparent to the healer is the aura. The healer may or may not be able to see the colors of the aura, but at some level the healer will be able to sense the energy of the person and, in particular, those areas of energy disharmony or disturbance indicative of disease.

The crystal is an excellent tool for use with mental or emotional aberrations, possessions, psychic attacks, etc., whether on the astral or mental plane. It lends itself to decongesting and balancing the various force centers or chakras of the body. It works especially well with the sacral centers and in cooling off prematurely stirred up and erratically moving Kundalini forces. As an all-around healing tool, it can even be used as an aid to break detrimental habits such as smoking and drinking.

For more detailed information regarding the results of a large number of experiments using crystals, see *The Crystal Book* by DaEl (Dale Walker).[3] The author also

---

[3] DaEl, *The Crystal Book* (Sunol, CA: The Crystal Company, 1983).

provides some excellent exercises for stimulating the healing force, which he invites others to try in their own experimentations. Two of them are reproduced here:

To Reduce Pain

Hold a crystal in your left hand. Place your right hand over the pain area. If the pain is on the right side, place the left hand with the crystal on the area, then grasp the left arm with the right hand. Hold these positions for half an hour. You can do this while watching your favorite television show.

Energy flows in the left hand and out the right hand. If you place your right hand over the pain area, amplifying the energy with the crystal, you will decrease the pain by sending energy into that area, which opens the blocked channels and promotes healing.

"The process of healing when viewed as energy, becomes simple," says DaEl. "When something blocks the flow of energy to the cells, the organs begin to die and you are sick. When energy is brought back to the area, new cells will form, rebuild the organs and you become well. . . . All anyone can do for someone else is to get energy to the cells so they will repair and heal themselves."[4]

The Aura Scan

The aura scan is an interesting healing technique offered by DaEl (Dale Walker). This is a basic technique demonstrated in many seminar lectures, but Dale has described it so interestingly I would like to quote it here from *The Crystal Book*. From personal experience, I can testify that it works:

---

[4]Source for this material: *The Crystal Book*, by DaEl (Sunol, CA: The Crystal Company, 1983), pp. 67-68. Copyright © by Dale Walker, 1983. Used by permission of the author.

1. Have the patient lie down on his back, hands beside him. Hold your crystal lightly between your thumb and the first three fingers about an inch away from the body. Start at the top of the head and work towards the feet.

2. When you feel heat or a tingling sensation, begin to move the crystal slowly over the body. As you move your crystal, be aware of any change in the feel of either the crystal or the body. The change may be a resistance to your movement, a sense of heat, a tingling sensation, coolness, or simply a feeling that something is there even if you have no sensation.

3. When you come to such a change, stop and begin making a circular counter-clockwise motion around the center of the disturbance. Keep making this motion until you feel the crystal get heavier, pulling your hand down toward the body. Stop the circular motion and touch the tip of the crystal to the body in the center of the circle.

4. Continue to move the crystal down over the front and sides of the body. Wherever you find a difference, make a correction using the circular, counter-clockwise motion.

5. When you reach the feet, go back to the top of the head and move the crystal from head to feet in a sweeping motion, as though you were smoothing feathers. This will align the auric energy pattern into a smooth even movement.

6. Turn the patient over and do the back side, using the same procedure. Where you find a difference, stop and make counter-clockwise circular movements over that area. Once again, smooth down the feathers.

7. You have the chakras very open, now. Close them down to normal by imagining a zipper at the feet. Reach down and grasp the zipper and pull it all the way to the top of the head.

8. When you finish, the person will be very relaxed, even sleepy. The whole body will be in balance. In a short time the client should feel rested and stronger. Aches and pains and tension will lessen and there will be a sense of well being.[5]

An interesting experiment to try regarding "zipping" discussed in section 7 above shows the power of this energy. Have someone stand with arm outstretched. Ask them to hold the arm stiff with as much strength as they can muster. Beginning at the base of the spine, pull an imaginery zipper up the back to the nape of the neck. It doesn't matter whether the person is actually touched by your hand or not. You can just move through the aura about two or three inches from the spine. Now try with all your strength to pull the outstretched arm down. You will find great resistance to your efforts. Now reverse the procedure, and "unzip" by pulling the imaginary zipper down the spine. Now try again to pull the outstretched arm down while the person tries with all his might to keep you from doing so. You will find that he cannot hold his arm up no matter how he tries. It moves easily down to your slightest touch. Always finish this experiment by pulling the zipper upwards again. Do not leave the person energyless.

Balancing the Chakras

The crystal pendulum is particularly responsive to chakra imbalances, and one can use the crystal pendulum to achieve rebalancing. Simply allow the crystal to swing freely in the chakra cone in either direction, using the intent of balance, until the swing stops. The chakra is then in balance.

---

[5]Source for this material: *The Crystal Book,* by DaEl (Sunol, CA: The Crystal Company, 1983), pp. 71-72. Copyright © by Dale Walker, 1983. Used by permission of the author.

Absent Healing

The use of a crystal as a focus of will and intention for absent healing, either by individuals or by groups, can be particularly powerful. In so doing the healer(s) enlists the aid of the elemental essence of the crystal or crystals to act as a sort of telephone line for the energies to be sent. The crystal supplements whatever technique the healer is already using. If color is being used, then send it through the crystal. If visualization is used, try to visualize the person as being inside the crystal with all of its harmonizing influences. Remember that visualization works best when the pain, injury or problem is ignored and the visualization is concentrated on the end product, the absence of pain or disharmony. See the patient as well, happy, joyful and full of the elixir of life and bouncing good health. You might try placing a crystal on top of a photograph of the patient, along with the appropriate visualization of this person as happy and well.

● ● ●

Only a few pioneers in crystal healing such as Dale Walker, Marcel Vogel and Ra Bonewitz have written about their experiments with crystals. All of these writers urge others to use their methods and to invent experiments of their own so that more might be known about the natural properties of these stones. Make notes—record your own experiments. If you discover something new, or create an experiment that can be duplicated or verified, others will want to know about it, too.

Chapter Five

Meditation with Crystals

Crystals have been used in religious practices for thousands of years. They have been consistently found in the tombs of priests and royalty in almost every archaeological dig. Stones from the quartz family are referred to over a thousand times in the Bible. Amethyst quartz was used as the twelfth stone placed in the foundation of the New Jerusalem, the ninth stone on the breastplate of Aaron, and is used in Orthodox Catholic Churches as a symbol of spiritual power. Every bishop has one. It is used to invoke the seven archangels for the altar.

There is a flow of eternal life energy that permeates all things visible and invisible, and all activities, human or otherwise. By using the crystal to amplify our tuning into this energy flow during meditation times, and being aware of remaining within its forcefield during waking hours, we can energize and enhance all inner intuitive awarenesses, as well as maintain health and balance and keep proper personal control over the manifestations of substance in our lives.

When praying, crystals promote greater clarity of thinking. Prayer is speaking to God through mind and heart, and meditation is listening to God through the inner senses and intuitive awareness. Our inner consciousness actually structures our reality according to the data it receives from both outer and inner senses. Meditation is the process of disconnecting the outer senses to enable the inner senses to be perceived.

Many researchers agree that the crystal was a holy object. In *The Quartz Crystal Story*, John Rea says, "Candle gazing through a crystal quartz suspended or resting at eye level for a few minutes before meditation or prayer was a practice common throughout the world, in such diverse areas as the Orient, the Near East, pre-Christian and Christian Europe and in the Americas as well. It was said to bring clarity and calmness to whatever endeavor it preceded, as long as it was used with good purpose, tempered by moral judgement and common sense."[1] In our present cultural climate, we exercise the outer senses far out of proportion to the inner senses. Meditation serves to equalize that proportion. We function best as unified beings when we are balanced and complete. Achieving a proper balance between the inner and outer senses creates harmony in mind, body and affairs.

If you already have a meditative practice which brings you a sense of balance and completeness, then continue on with what works best for you, but hold a crystal in your left hand. You may also place a crystal upright in front of you. You'll find your meditation time enhanced with unusual and informative experiences. If you have no present method of meditation, here are some simple ones:

Affirmation

Sit with spine straight, shoes off and feet flat on the floor. With your crystal in your left hand and the point facing you, close your eyes and repeat over and over: "I am the Light of God," or "I am a radiant Being of Light temporarily

---

[1]John D. Rea, *The Quartz Crystal Story*, a pamphlet published by High Peak Crystal Co., Inc., 1272 Bear Mt. Ct., Boulder, CO 80303. Used by permission.

using a physical body," or "I am filled with the Light of the Christ." Using these or other affirmations such as "I am Love," "I am Wisdom," "I am Health," etc., will strengthen all of these qualities within you. Since the crystal amplifies whatever energy you are projecting, you'll find these qualities being amplified within yourself.

Your Newly Acquired Crystal

When you first receive a crystal, no matter how you get it, whether it is given to you, purchased or found, take it into meditation with you and meditate on the reasons why it has come to you. Ask the crystal what uses it is in harmony with and what it would like to be used for. The response may be an intuitive feeling, a vision of a particular use, or even a sense of a verbal reply. Just leave your mind open to whatever happens, as you contact the living intelligence behind the physical form, the essence or elemental energy of the crystal.

Crystal and Candle

One of the meditations recommended by John Rea is as follows: "Suspend a crystal at eye level from a string or by setting the crystal in a cup of something that will not ground it, such as rice or beans (not salt, sugar or sand— these will ground the crystal and restrict its aura). Keep the crystal upright, if possible. Place a lighted candle behind the crystal so that you are looking through the crystal at the flame, in a meditation or praying position."[2]

Observe the flame through the candle and as your brain relaxes more deeply into a meditative state watch the striations of light as they reflect through the crystal. Be aware of your breathing and of how your breath goes in and out of your body. Allow thoughts to come and go in

---

[2]John D. Rea, *Notes on the Uses of Quartz Crystal*, a pamphlet published by the High Peak Crystal Co., Inc., 1272 Bear Mt. Ct., Boulder, CO 80303. Used by permission.

your mind, but if they become the focus of your attention instead of the candle flame as seen through the crystal, gently bring your attention back to the breathing and the candle. Practice once a day for fifteen to twenty minutes.

Get with the Flow

Holding a crystal in your left hand with the point towards your head, lie on the floor or bed with your head to the north, so that you are in line with the polar energies of the earth. As your body and brain slow down, feel the pulsing of the life force in your body. Submerge yourself into it. Flow into this pulsating, living part of the universe. Feel as if you are being swept off head-first, still in a prone position, and floating into a sea of energy. Mentally float off into the stratosphere, with the earth becoming smaller and smaller as it recedes from your inner view. Go with the tide, gently move along in outer space in a spiral motion, and be completely immersed in the cosmic rhythms, feeling yourself a part of the Divinity.

When finished, infill with the protective White Light, while slowly returning to full consciousness. The White Light represents a universal protection created by the mind, using man's natural devotional nature. Simply visualize a brilliant Light all around you, know that the Light is God and that God is protecting you from all forms of negation.

Meditation on the Inner Way

Sit with spine straight, holding a crystal in the left hand. Use any mantra you wish to reach a deep reflective state. As you find yourself turning inward, substitute the pronoun "I," or the phrase "I am," for whatever mantra you have been using. You'll find the inner perceptions turning upwards, instead of inwards or down.

Reflect on the concept of "I" as expressing your total Being which has both sensory and spiritual awareness.

Think of your thoughts as branches or twigs floating on a stream. Allow the ideas associated with the concept of "I" to just flow. Don't grasp hold of any of them. Don't allow the conscious mind to build on any particular association. Just see and acknowledge the relationship, then let it go. Come back to the original concept and repeat. With each new association that rises, let it go, and come back to the original concept.

After a time return gradually to waking awareness. Fill your body with the White Light of protection, recognizing this Light as the Oneness with all that IS, a Light that will transform your life in many important ways.

Rainbow Meditation

Sit in a comfortable chair, spine straight, holding a crystal in your left hand. Imagine at your feet a bright, glowing ball of violet color. See it as the pulsating, vibrant violet energy of high spirituality, representing one who is searching for the deeper meanings of life and existence.

Let this ball expand and rise until it fills and permeates every atom of your body, drawing the Light energy upward until it comes out through the top of your head. Then see it explode with a flashing cascade of violet fire and flow down over you like an energizing fountain.

Repeat the procedure with the rich deep tones of indigo, for these colors represent the searcher who truly seeks his or her own purpose in Life.

Follow this with the same procedure using other colors: blue, which brings the feeling of Divine Guidance; green, reflecting great joy, anticipation and healing; yellow, representing deep thought or intellectual activity; orange, representing wisdom, justice, creativity, and good will to all; and bright, clear, scintillating red which stands for vital, optimistic zest for life.

Try to visualize clear, bright colors but let your mind be free to bring in whatever associations it wishes as you draw each ball of color up through your body to explode in a shimmering cascade of color and energy down over your whole Being.

Do this exercise for three days. Then reverse the procedure, going from red to violet. The first exercise will seem to draw your attention down and in; the second will draw your attention upward towards higher spiritual attunement.

Group Meditation with a Crystal

A single large crystal, or a large cluster of crystals can be used in group meditation. Form the members of the group in a circle around the crystals and focus your attention on the stone. Before meditation, it is good to discuss the precise intention of the group, what it is that is expected to be accomplished, then the energy of all, including that of the crystal can be focused as clearly and precisely as possible. If every person in the group has a different idea or intent, the opposing energies may just cancel themselves out.

To insure against conscious or unconscious misuse of energies from the crystal, the group should be aware of the necessity to keep the crystal cleansed. This could be done by appointing someone to do it regularly, or it could be done by the group as a whole before beginning meditation. (See Chapter Six on "Care of Crystals.") Frequent cleansing of the crystal helps not only to maintain the programmed positive purpose of the desired intentions of the group, but also to reinforce them.

The following meditations are not for beginners. They focus very powerful energies, and are definitely not a place for the unaware and the unawakened to experiment.

Merging with the Crystal*

Look deeply into the heart of the crystal. Close your eyes and imagine it growing larger until it is bigger and more powerful than any crystal in the world. Feel the energy coming from it, and imagine those great energies enveloping it and you.

Now imagine your crystal growing even larger, until it is as big as a house. Go inside and look around. See and touch the floors, the walls, the ceiling. Can you look out through the crystal? If so, what do you see, feel, smell? Use all your senses while you are inside your crystal house. How are sight, sound, touch, taste, and smell affected?

Now that you are one with your crystal, imagine you are getting bigger and bigger; your crystal and you with it are getting bigger and bigger, until everything you see and can imagine is inside the crystal: the surface of the earth, the sky, the sun, moon and stars, all the universe, all of space and time. Imagine the crystal is so big that it has no walls or edges—everything is crystal. Now imagine that it is moving, that it is a great moving river of crystal beyond time and space. Feel the river of crystal moving through you, cleansing all pain and suffering, all sorrow and loss, cleaning out all dark spaces. Feel yourself merging with this gently shining, moving sea of crystal like a river of gentle, cleansing white light that glows and shines in everything, moving through and permeating all things.

After a time, return to full waking consciousness. Once merged in this way with your crystal you can return to this feeling at any time during the day. As with all things, it becomes easier with practice.

CAUTION: If you are a beginner using crystal energies, it is a good idea to meditate on the crystal itself, rather than try to go inside the crystal. The more you

---

*This meditation has been adapted from John D. Rea's *Notes on the Uses of Quartz Crystal,* a pamphlet published by the High Peak Crystal Co., 1272 Bear Mt. Ct., Boulder, CO 80303. Used by permission.

become intimately involved with the energy structure of the crystal by going into it, the closer you contact the elemental energy—that essence of being that is the crystal itself, the crystal deva, as it were—and the more difficult it is to remain detached when assessing information from the meditation. Let your own sense of rightness and wrongness be your guide, maintaining an awareness that fantasy is always present and must be differentiated from reality.

Pyramid and Crystal

A crystal placed at either the apex of a meditation pyramid or at the energy focus can be very powerful. The energy of the crystal combined with the energy of the pyramid requires a great deal of conscious high attunement, and a great deal of clarity of purpose to control it correctly. Don't try this one until you have reached a state of attunement where you are living on at least a mental level rather than an emotional one. On the other hand, one can enhance the energy levels of the crystal itself, to "charge it," by placing it within a pyramid shape for a period of time, say, overnight.

● ● ●

There are as many meditation practices as there are people to think them up. Try adding crystals to your favorite meditations. Hold them in your hand, or on your lap. Gaze into the depths of a crystal on floor or table. Sit inside a circle of crystals formed in the shape of the Zodiac, or a Star of David. Multiply the force of your thought energy with the amplification of the crystals. Write down your experiences, so that others might share them.

Chapter Six

Care of Crystals

When you first buy a crystal, scrub it with an old brush using warm soapy water. This will remove any foreign particles from the surface and crevices. This will also remove any oxalic acid residues that may cling to the stone. Oxalic acid is a poison used to dissolve red iron oxide and other minerals from the newly mined quartz, and it is generally harmless in small doses. However, the acid can accumulate in the body and should be handled with extreme caution according to John Rea of the High Peak Crystal Company. Many crystal users recommend cleaning the crystal of any previous vibrations by burying it in sea salt. Rea also recommends that you avoid using rock salt or any sea salt that has aluminum in it as that defeats your purpose. Other kinds of baths to clean the crystal have been used over the centuries—spiritual ceremonies, sunlight, water cleansings and meditation have all been used in addition to the sea salt technique. If you do use sea salt, avoid any kind of aluminum container for the cleansing process; glass, china or porcelain make ideal containers for this ritual. Some researchers recommend that you bury your crystal in sea salt overnight, or for three or seven days. Perhaps the length of time depends on the sensitivity of the user. You will have to find the method that works best for you. Just make sure that you throw away the sea salt after each use.

You can also immerse your crystal in warm salt water for ten minutes to rid it of any negativity it may have acquired. When the stone is dry, expose it to unobstructed sun's rays for an hour or two, which will begin to charge the crystal for you, and with the willed intention of transforming any negative energies into positive. Allow your own sensitivities to be the judge as to how long the crystal should remain in the sun.

Crystals are used primarily in dealing with subtle energies which tend to have an effect on the very nature of the matter that makes up the crystals; therefore it is advisable to cleanse your stones occasionally, more frequently, if they are used daily such as in healing.

Techniques for cleansing vary greatly from user to user, and through experimentation you can find the techniques that feel right for you.

Water is a universal cleanser and some people prefer to wash their crystals in naturally flowing water, such as creek or river. This might be more difficult for city folk or desert dwellers, so flowing tap water can be substituted. Do not use hot water on your crystals, as it could shatter them. Use cool to warm water only. During the washing process use your mind to will or intend that all undesired energies will be washed away and that all desirable energies should remain.

An equally rapid and effective method of cleansing is the visualization method. Simply hold the crystal in your hands and visualize positive energy flowing through the crystal, sweeping away all negative or undesirable energies. You might visualize a stream of light and love flowing through the crystal, sweeping away all impurities, and transforming them into positive healing energy.

It is the responsibility and prime objective of all workers for the Light to cleanse Earth's energies while working on their own spiritual upliftment. When cleansing the crystal of whatever negative energies it may have picked up during the course of its use, one should always take care to transform this energy into other, more positive energies, not thoughtlessly scatter it abroad upon the face of the globe. Unless it is transformed, the negative energy will simply go elsewhere, and attach itself to someone else. We must take time to nullify some of the karmic load of negative energy now present on our planet, or we will never raise the consciousness of mankind as a whole.

On the subject of cleansing, one's own being should be cleansed before and between patients, if one is doing healing in any form, crystal or otherwise, so that any personal negative energies, or those inadvertently picked up from a previous patient will not be passed on to the next.

Let me share with you a remarkable personal experience I had during a psychic seminar in Phoenix several years ago. It left me awe-stricken and was an experience I will not likely soon forget.

Sunday morning activities of the seminar were devoted to spiritual healing in which those who were actively engaged in spiritual healing formed a circle at the front of the room around the person to receive healing energies. The rest of us sat in our seats and, with palms outturned towards the circle, added our own bit of healing energies.

Although I had been working many years with Light energy and spiritual growth, I had not, up to that point, focused much of my attention in the physical healing area. As I raised my hands, palm outwards, toward the circle of healers, there was suddenly a clairaudient voice in my mind which said sharply, "Don't do that!"

Startled, I dropped my hands, and responded (in my mind, of course, for an outward response would surely have turned heads in my direction), "What do you mean?"

The voice replied as clearly as before, "Never turn healing energies to someone else without cleansing your own negative energies first, or you'll just add your own negation to theirs."

Chastened, I asked, "What should I do?"

The voice continued, "First cleanse yourself with the White Light." I did as I was told. "Now fill yourself with blue Light." This I did, and the voice went on, "Now fill yourself with green Light." When I had accomplished this, the voice said, "Now you can send healing energies," and as a parting shot the voice left me with the admonition, "And never, ever do that again!"

Amused at the audacity of the last injunction, but nevertheless grateful, I sent a mental thank you to the departing spirit (what else could it have been?) and returned to my task, much amazed that someone (or something) had cared enough to interrupt and enlighten me. I've never failed to pass this bit of information on whenever there was an opportunity throughout the ensuing years. The amplification properties of the crystal can magnify our thoughts and energies either to our harm or our good.

Some users feel that personal crystals, that is, crystals that are used solely by one person, are best kept out of sight and not handled by others, so that the large amount of personal energy involved with such a crystal is not disrupted. It is my feeling that this should not cause that much concern. If such occurs, and you sense any negativity at all, cleansing is as simple as any of the methods described. Dip it in salt water, and all negativity is removed instantly. Remember to make your intent in using the crystal one of purity, wholeness, love and healing and to consciously, willfully transform all negative energies from whatever source into love and light, while willing that all previously programmed positive energies remain intact. Such conscious intent will keep your crystal free and clear of negative influences at all times.

In order for the crystal to acquire your vibrations, let it be a companion. When you meditate, hold it in your hand. If that is not comfortable, place it near or on top of your body so that it is within your auric field and will pick up your vibrations. Wherever you go, take it with you. Keep it on your person. The longer it is in your presence and within your vibrations, the sooner it will become an extension of yourself, and can be used as such.

Bibliography

Achad, Frater. *Crystal Vision Through Crystal Gazing.* Jacksonville, FL: Yoga Publication Society, 1923.

Alper, the Reverend Dr. Frank. *Exploring Atlantis,* Vols. I and II. Phoenix, AZ: Arizona Metaphysical Society, 1982.

Baer, Randall N., and Vicki B. *Windows of Light.* New York: Harper & Row Publishers, Inc., 1984.

Bonewitz, Ra. *Cosmic Crystals.* Wellingborough, England: Turnstone Press, 1983.

Cayce, Edgar. *Scientific Properties and Occult Aspects of 22 Gems, Stones and Metals,* from the Edgar Cayce Readings. Virginia Beach, VA: ARE Press.

Crow, W.B.: *Precious Stones: Their Occult Power and Hidden Significance.* York Beach, ME: Samuel Weiser, Inc., 1968.

DaEl. *The Crystal Book.* Sunol, CA: The Crystal Company, 1983.

Deaver, Korra L. *Rock Crystal: Nature's Perfect Talisman.* Little Rock, AR: Parapsychology Center, 1978.

Evans, Joan. *Magical Jewels of the Middle Ages and the Renaissance.* New York: Dover Press, 1976.

Fernie, William Thomas. *Occult and Curative Powers of Precious Stones.* New York: Harper & Row Publishers, Inc., 1907.

Ferguson, Sibyl. *The Crystal Ball.* York Beach, ME: Samuel Weiser, Inc. 1979.

Finch, Elizabeth. *The Psychic Value of Gemstones.* Cottonwood, AZ: Esoteric Publications, 1979.

Glick, Joel, and Julia Lorusso. *Healing Stoned.* Albuquerque, NM: Brotherhood of Life Books, 1979.

Hodges, Doris. *Healing Stones.* Perry, IA: Pyramid Publishers of Iowa, 1961.

Jaegers, Bevy. *The Art of Crystal Gazing.* Sappington, MO: Aries Productions, 1983. If you can't find the book in your local bookstore, order from: Aries Productions, PO Box 29396, Sappington, MO, 63126. $2.90 ppd.

Kunz, G.F. *Curious Lore of Precious Stones.* Philadelphia: J. B. Lippincott, 1913; reprinted in 1970 by Dover Press, New York.

Mella, Dorothee. *The Legendary and Practical Use of Gems and Stones.* Albuquerque, NM: Domel Press, 1976.

_____ *Stone Power.* Albuquerque, NM: Domel Press, 1979.

Melville, J. *Crystal Gazing and Clairvoyance.* York Beach, ME: Samuel Weiser, Inc., 1974.

Rea, John D. *Notes on the Uses of Quartz Crystal.* Boulder, CO: High Peak Crystal Co., 1982.

_____ *The Quartz Crystal Story.* Boulder, CO: The High Peak Crystal Co., 1982.

Richardson, Wallace G., as channeled through Lenora Huet. *Spiritual Value of Gemstones.* Marina del Rey, CA: DeVorss Publishing Co., 1980.

Stewart, Nelson. *Gemstones of the Seven Rays.* Madras, India: The Theosophical Publishing House, 1939; reprinted in 1975 by Health Research, Mokelume Hill, CA.

Thomas, William and Kate Panitt. *The Book of Talismans, Amulets and Zodiacal Gems.* N. Hollywood, CA: Wilshire Book Co., 1970.

Wright, Ruth, and Robert Chadbourne. *Gems and Minerals of the Bible.* New Canaan, CT: Keats Publishing, 1977.

My Crystal Journal

MY CRYSTAL JOURNAL

MY CRYSTAL JOURNAL

MY CRYSTAL JOURNAL

MY CRYSTAL JOURNAL

MY CRYSTAL JOURNAL

MY CRYSTAL JOURNAL